地球家园

小猪爱可的动物保护行动

[美] 丽莎·弗兰奇 著　　[美] 巴瑞·戈特 绘　　张玉亮 译

江西科学技术出版社

Original title: ECO-Pig's Animal Protection: A Planet Called Home

Copyright © 2010 by Abdo Consulting Group, Inc. International copyrights reserved in all countries.

All rightsreserved. No part of this book may be reproduced in any form without written permission from the publisher.

The simplified Chinese translation rights arranged through Rightol Media

（本书中文简体版权经由锐拓传媒取得Email:copyright@rightol.com）

版权合同登记号 / 14-2016-0011

图书在版编目（CIP）数据

地球家园 : 小猪爱可的动物保护行动 : 英汉对照 / (美)弗兰奇著；(美)戈特绘；张玉亮译.
-- 南昌 :江西科学技术出版社, 2016.8
（小猪爱可讲环保）
ISBN 978-7-5390-5488-9

Ⅰ.①地… Ⅱ.①弗… ②戈… ③张… Ⅲ.①环境保护 – 少儿读物 – 英、汉 Ⅳ.①X-49

中国版本图书馆CIP数据核字(2016)第026822号

国际互联网（Internet）地址：http://www.jxkjcbs.com
选题序号：KX2016083   图书代码：D16001-101

小猪爱可讲环保
地球家园：小猪爱可的动物保护行动

文 / (美) 丽莎·弗兰奇   图 / (美)巴瑞·戈特   译 / 张玉亮
责任编辑 / 刘丽婷   美术编辑 / 刘小萍   曹弟姐
出版发行 / 江西科学技术出版社
社址 / 南昌市蓼洲街2号附1号   邮编 / 330009
电话 / (0791)86623491   86639342(传真)
印刷 / 江西华奥印务有限责任公司
经销 / 各地新华书店
成品尺寸 / 235mm×205mm   1/16
字数 / 50千   印张 / 8
版次 / 2016年8月第1版   2016年8月第1次印刷
书号 / ISBN 978-7-5390-5488-9
定价 / 50.00元（全4册）

赣版权登字-03-2016-7   版权所有，侵权必究
（赣科版图书凡属印装错误，可向承印厂调换）

Eco-Pig pedaled
through the town of To-Be.
His two very best friends joined him
on a bicycle for three.

　　小猪爱可和他最要好的两个朋友踩着一辆三人自行车，在未来小镇上愉快地穿梭。

They rolled past the wind farm
and the Green Goodies Shop.
But they stepped on the brakes
at the electric bus stop.

他们先是经过风力发电厂，又渐渐
地将绿色小铺甩在了身后……突然，他
们在电动巴士站前刹住了车。

E.P. stood and he stared.
So did Lou and McGee.
E.P. asked his two friends,
"Do you see what I see?"

　　小猪爱可终于站稳了脚跟，他被眼前的景象惊呆了，洛尔和麦吉也愣住了。小猪爱可有点儿不敢相信自己的眼睛，迟疑地问自己的两个好朋友："你们也看到了吗？"

There was a family of polar bears,
a black-footed ferret or two,
a sea otter, and a gray wolf.
"Where did you come from?" cried Lou.

　　站在他们面前的是北极熊一家、一只黑足
雪貂、一只海獭和一头灰狼。
　　"亲爱的朋友们，你们从哪里来啊？"洛
尔冲他们大声喊道。

"We came on the bus," said Pat Polar.
"We've been riding all day.
We've all lost our homes.
We need someplace to stay!"

　　"我们是乘公交车来的，"北极熊帕特回答说，"我们已经在车上折腾了一天了。我们失去了自己的家园，想要找一个落脚的地方！"

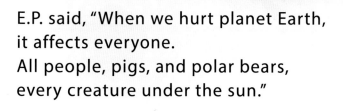

E.P. said, "When we hurt planet Earth,
it affects everyone.
All people, pigs, and polar bears,
every creature under the sun."

小猪爱可叹了一口气说："当我们对地球造成了伤害，生活在地球上的每个生物都会受到影响。所有的人类、猪和北极熊，世界上的所有生物都在劫难逃。"

"We lived on a prairie," sighed Fred Ferret.
"What a great place to roam!
Since it's turned into farmland,
we have no place to call home!"

　　"我们以前生活在大草原上，"雪貂
弗莱德叹息道，"漫步在草原上，是件
多么美好的事儿啊！但随着草原慢慢变
为农田，我们也就无家可归了！"

9

"And the sea ice is melting," said Pat,
"out from under our feet.
That makes fishing too hard!
Now we have nothing to eat!"

　　"海上浮冰也在慢慢融化，"北
极熊帕特说，"我们失去了立足之
地。鱼也越来越难捕捉了！现在，已
经没有东西能让我们填饱肚子了！"

Olive Otter added, "I floated
right into a big oil spill.
I tell you that water pollution
makes this poor otter ill!"

海獭奥利夫补充道："我游进了一大
片浮油层中。跟你这么说吧，就是可恶的
水污染让我这只可怜的海獭病倒了！"

"I'm George," said the gray wolf.
"I need a place to run free.
Without a wide-open space,
it's so hard to be me!"

　　"我叫乔治，"灰狼自我介绍道，"我需要一个能够自由奔跑的地方。如果没有开阔的原野，我很难展现自己真实的一面——与生俱来的狼性！"

"With no place to live," said E.P.,
"wolves could be gone forever!
We can't let that happen,
we all have to say never!

"没有了立足之地，"小猪爱可说，"狼就会绝迹的！我们可不能让这种情况发生。对于动物灭绝这样的事情，所有人都会坚决反对。

"If I came back from recycling,
and my tree was not there,
and my apples were gone,
that sure wouldn't be fair.

"如果哪天我完成垃圾回收工作，
回到家却发现我的苹果树不见了踪影，
我的苹果更是不翼而飞的话，那对我可
就太不公平了。

"Every creature on Earth should have a good life. That includes Fred Ferret and Betty, his wife.

"地球上的每一种生物都应该过上优质美好的生活。雪貂弗莱德和他的妻子贝蒂也不例外。

"Every sea otter,
every great polar bear,
and George the gray wolf
all need us to care!

"每一只海獭，每一头大北极熊，还有像乔治一样的大灰狼都需要我们的关怀。

"We all can have homes
and enough food to eat,
if we look out for each other.
Wouldn't that be neat!?"

"如果我们相互关照，我们就会有温暖
的家和充足的食物。这种感觉多温馨啊！"

"Let's turn down the heat," said Pat.
"What a great way to start!
If you help stop global warming,
it shows you have heart!"

　　"让我们一起行动起来'关掉'地球
的'暖气'吧，"北极熊帕特说，"现在
就是一个很好的开始！如果你愿意参与阻
止全球变暖，那表明你是个有心人！"

冷

热

学校

You can walk, ride a bike,
or take a bus to your school.
The fewer cars on the road,
the more Earth stays cool!

你可以步行、骑车或乘坐公
交车上学。路上的汽车越少，我们
的地球就会变得越凉爽！

19

"I have a favor to ask," Olive said.
"Please don't pollute water!
Remember somebody lives there,
and it just might be an otter!

　　"我有一个小小的请求，"海獭奥利夫说，"请大家不要污染水资源！请一定要记得那里还有很多生物居住着，其中很有可能就有像我这样的海獭哦！

"When you go for a swim,
don't leave trash on the beach.
Keep soda cans, plastic bottles,
and plastic bags out of reach."

"当你游泳时，请不要把垃圾扔在沙滩上。请一定要把饮料罐、塑料瓶和塑料袋扔进垃圾箱，让这些垃圾远离沙滩。"

21

"And if you obey special laws that make our homes a safe place," said George, "I'll howl at the moon with a grin on my face!"

"如果你能遵守相应的环保法规，那将会使我们的地球家园成为安全舒适的地方，"灰狼乔治说，"我将面带笑容，望月长啸！"

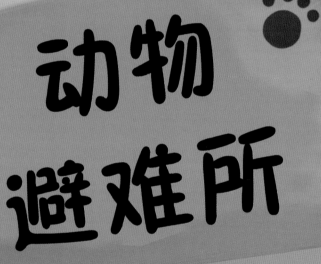

动物
避难所

"Wherever we are," E.P. said,
"there are creatures nearby.
We have friends on the land,
in the sea, and in the sky!

"无论我们在什么地方，"小猪爱
可说，"周围都有生物存在。陆地上、
大海里、天空中都有我们的朋友！

23

"Let's try to remember
that each thing we do
could save a friend's life.
Believe me, it's true!

"我们要记住，我们所做的每件事都
可能拯救一位朋友的生命。相信我，这绝
对是真的！

停

"Learn as much as you can
about each habitat.
Help guard them from danger.
I'd really like that!"

"尽可能地多了解每一种生物的栖息地。确保它们能够远离危险。我是真的非常喜欢和支持这种做法！"

"Excuse me, kind sir,"
Fred Ferret said to E.P.
"If we're quiet and neat,
may we live in To-Be?"

"打扰一下，仁慈的先生，"雪貂弗莱德对小猪爱可说，"如果我们保持安静、保证清洁卫生，我们可以在未来小镇居住吗？"

"Of course you can live here,
but first you should eat!
Let's all go to the diner!
I'm buying, my treat."

"你们当然可以生活在这里。但是首先，你们得吃点东西啊！我们一起去吃饭吧！今天我请客。"

27

Everyone got fresh fish.
They shared green peppers and peas.
When offered green apple pie,
they all nodded and said, "Please!"

　　每一位朋友都吃到了新鲜的鱼。他们还分享了青椒和豌豆。当青苹果派端上来的时候，他们不约而同地点了点头，齐声说："请一起分享吧！"

Olive smiled and said, "Thank you, E.P.
What a great day!
Now To-Be is our home,
a friendly, safe place to stay!"

　　海獭奥利夫微笑着说："谢谢你，小猪爱可。今天真是太棒了！现在未来小镇就是我们的家了。这是一个充满友爱和安全感的家园！"

必学词汇

生态学——研究动植物与它们所处环境间关系的一门科学。

电动公交车——由不会污染空气的清洁电能驱动的公交车。

全球变暖——地球温度逐渐增高，进一步引发气候变化。

栖息地——动物通常生活和成长的地方。

浮油——从油轮中渗漏到海中的石油。

污染——用人造废弃物破坏环境的行为。

回收利用——将废物、玻璃或易拉罐分类回收，以便再次利用。

风力发电厂——将风能转化为电能的工厂。

你知道吗？

- 地球每年都有2.7万种动物灭绝。

- 地球有四分之一的动物濒临灭绝。

- 很多物种因为它们的栖息地消失，同时也因为它们无法转移到新的环境而灭绝。

- （生物）栖息地遭到破坏的主要原因是：

  a.荒地变成农田（垦荒）。

  b.污染破坏了生态系统和栖息地。

  c.燃烧化石燃料造成全球变暖、冰雪融化、海平面上升。

  d.人类为获取木材和燃料而砍伐森林。

- 所有生物都是生物链的一部分。

- 如果一种动物灭绝，可能我们不会注意到，但这却改变了整个地球的生态系统。

打造美好动物家园的
更多方法

与你的父母聊聊
你们在家中可以做些什么

1. 尽量多了解关于濒危动物的知识。

2. 尊重野生动植物。不要打扰你家附近的巢穴或自然栖息地。

3. 在家附近安装鸟浴池和喂鸟器。

4. 种植花草、乔木和灌木来吸引鸟类、蜜蜂和蝴蝶。

5. 不要往排水沟里或大街上乱倒垃圾。垃圾会流到河流、湖泊和溪流中去，造成污染。

6. 捡起沙滩上的塑料袋。乌龟、海豚、海獭和其他海洋生物吞下塑料袋后会死亡。

7. 剪碎包装汽水和果汁易拉罐的塑料包装。动物会误吞下它或者被它卡住喉咙。